C0-AFV-371

A ribbon of snow that wraps a package of ice, the Linn Cove Viaduct flows like a frozen current around Grandfather Mountain. Ironically, these most breathtaking miles of the Blue Ridge Parkway were the last to be completed. If the parkway is the crowning glory of the mountains, then the Linn Cove Viaduct, especially in winter, is the jewel in the crown.

Whitewater Falls, southeast of Cashiers near the South Carolina border, seethes and rampages in one of the most dramatic descents in the eastern United States. Crashing more than 400 feet from top to bottom, the upper cascade of the Whitewater River is deafening.

There is a certain "rightness" about the farms which are visible from the Blue Ridge Parkway such as this one near the Virginia line. They nestle within the few farmable acres of

North Carolina's mountains are their own gift; you must be ready to receive what is offered at any hour on any day. All of this is yours; you can take what your eyes see and your soul embraces. This particular arrangement of crests is in the vicinity of Waynesville.

ISBN: 0–942399–02–1

Library of Congress Catalog Card Number: 88–80530

Hardcover ISBN: 0–942399–02–1

Printed in Singapore through Palace Press.

DAWN'S EARLY LIGHT

NORTH CAROLINA'S ETERNAL MOUNTAINS

Photography by Chip Henderson, Steve Muir and Scott Larson
of Henderson/Muir Photography

Text and Captions by Glenn Morris

Design by Russell Avery, Avery Designs

Published by Lightworks, Raleigh, N.C.

The mountains are closer to being a continuous surprise package than any other region in North Carolina. Imagine the land itself: Little is straight here, and less still is level. These conditions infuse the hills with an abundance of serendipity. You never know what to expect around every bend except an unanticipated delight for the senses.

A gentle breeze may surround you with the sweet, fresh smell of a stream and the slap-plop of its water's flow and, at another season some wind, somewhere in the hills, could bring the wood-smoke smell of a warming mountain cabin. As sure as the road you follow leads away, you'll leave what touches you behind for the unfolding of a hillside of Holsteins or the breathtaking splendor of roadside rhododendron.

This is how you experience the mountains, vignette after vignette, one photograph-in-your-mind's-eye after another. In fall it seems, there is more than the film of memory can ever record.

The unpredictable allure of the mountains pulls at something deep within you like a calling. You don't visit the mountains; you make a pilgrimage that has as its destination the journey itself. The discoveries made during the passage settle a yearning for peace and serenity. This unwrapping of a journey-for-the-senses is as comforting as snuggling into a familiar bed on a chilling night—the warmth you feel is far greater than that provided because you are warming your soul.

These hills harbor as many moods to experience as they hold views for the looking. Their diversity is not consumable at once, therefore you must return year after year to taste it, a sip at a time until you have a sense of its full flavor. The thundering explosion of Whitewater Falls abounds in awesome energy, the sleepy vales of the New River country radiate a dignity of diligent stewardship, and the towers of the Great Smokies slumber with a brooding power. Each magnificent vista separates further into smaller composites that project their own message. For every waterfall there is a languid pool, for every rolling farm-

Along the Oconaluftee River near Cherokee, gold is abundant. Like every metal nugget coaxed from the earth which has a value and look all its own, every October in the mountains glows with an individual richness that is forever matchless.

stead, a staggeringly-steep mountainside, and within the folds of giant peaks are the embroideries of its wildflowers. There is not enough time in life to see it all, so don't hurry: savor what you do see to the fullest.

What is known about the mountains can help you grasp what you feel when you are there. The easiest things to know are those which can be measured. Within the boundaries of North Carolina are Whitewater Falls, considered to be the highest waterfall in the East; nearby Whiteside Mountain, with the highest cliff in the East; Mount Mitchell, the highest mountain in the East; the New River, the second oldest river in the world (the Nile outdates it); and Great Smoky Mountains National Park, the most visited park in the nation. There are several other items of note: for example, the Great Smoky Mountains have more species of trees naturally than Europe; the spruce and fir forests of the higher elevations are normally found 900 miles to the north. These are the things man has found in the mountains and could apply yardsticks to. Yet the origin of the very mountains themselves weaves a trail as circuitous as most farm-to-market roads herein.

The history of the mountains almost exceeds your capacity for comprehension. The story is not only complex and in many respects incomplete, but it occurs (mountain-shaping forces are still active) in a time frame that shatters human reference. You experience in your lifetime only a minuscule instant of the mountain's geological story. In many respects, a journey through the high country is a passage into the very ancient past.

Take one familiar destination such as Grandfather Mountain. Just like some paternal family patriarch who cannot recall his date of birth, Grandfather Mountain holds a key to the history of his sprawling extended family. Geologists know this ancient summit as a *window*. Peer from the swinging bridge on a clear day and you are peeking into the basement of the Blue Ridge, looking at foundation rocks easily 600 million years old, probably more. It took a mere 300 million years for erosion to wipe the glass-clean film off the window permitting this insight.

The sheer power of a waterfall is humbling but the realization of what that power is accomplishing invokes awe. A waterfall is but a moment in geologic time. Even as you watch, the fall alters its course, but not in any measurable amount.

If such durations of time boggle the mind consider this additional astonishing fact: What is now Grandfather Mountain began its geological history *at least* 150 miles east of its present location. Like most of his hilly family, Grandfather was shoved into place, sliding across other rock formations en route. What moves mountains? Collisions of *BIG* items such as continents generate the force to slide miles-thick layers of rock across one another like so many stacked layers of corrugated cardboard.

Geologists are in general agreement that the salient event shaping the mountains occurred 300 million years ago when the ancient African continent collided with North America. The land rumpled in response. Great sheets of rock were shoved, folded and thrust upward over existing layers in a massive northeast-to-southwest-trending arc. Imagine pushing one end of a rug against a wall: somewhere the rug will wrinkle and fold. Apply enough force to rock—ram one continent with another—and the same results appear.

But if you had witnessed this earth-shattering event—which took countless millennia to complete—you would not see the Great Smoky Mountains, nor the Black Mountains, nor the Snowbirds. Instead, you would see youthful, just-formed ranges far loftier than today. The ridges and slopes of today's mountains—wrinkles in earth's ancient throw rug—were deeply underneath the then-present surface, perhaps as much as nine miles below.

Crystal by crystal, the process of erosion peeled away the covering. Water and wind transported the wrapping to ancient seas. Thus carried, these removed particles settled to form the sediments of the coastal plain, revealing the once-buried wrinkles as the mountains of today. What we measure as different heights of the mountains is essentially the difference in hardness of the inherent rockstuff. The more erosion resisting the rock, the higher the mountains we see. The fact that mountains made of very hard rock, such as the quartzite pilings of Pilot Mountain, are not as tall as the Great Smoky peaks, is evidence of longer erosion—much longer.

The rending and rifting history of the mountains is still being unknotted by geologists. Each revelation of the cloudy distant past only adds to the appeal of the present's haze-shrouded hills. As sure as the vales harbor ancient mysteries, they also sequester an almost secretive, slightly guarded way of life. The people of the mountains are as protective and reticent about their lives as the very hills are unspeaking and locked.

If landforms can shape an attitude of life, then *live and let be* would be the guiding rule of mountain natives. Why? Because the isolation imposed by the physical barriers of the mountains demanded self-reliance. Even in this modern era in which roads and communications have shortened distances and the time required for daily chores, the mountaineer remains resolutely independent. He would rather do for himself, for that way he knows things will be done to his satisfaction. He rarely does more than is necessary to accomplish his ends for to do so would waste a portion of his most treasured commodity—time.

There is little extravagance among mountain natives for the simple reason that there is little time for it, and little money as well. Living merely accomplishing all that is required for food, clothing, and shelter consumes most hours of the day. It is hard living. What is left over in time offers a small chance to get ahead of the race or relax. True to a region of extremes, no one plays like the mountain natives, for few work as hard at just living. The free-wheeling jubilation of clogging, and the vigorous beckoning of a banjo's picking portray the other half of the living equation: If you work hard, you can play hard, too.

Exuberance balances diligence here. The mountaineers explosion at play is like the eruption of autumn color on a hillside—reward for a season's hard work. Spectators are welcome and the show will rub off on you as well.

All of this is for the taking—a communion of sorts. Taste the clean waters, inhale deeply the crisp air, open your eyes and let them be a window for your soul. Peer 'round every bend, open and receptive, and the hills shall bequeath an eternal gift. Make the most of what you receive, for it is yours for a lifetime. These mountains are yours; yet they belong to no one and to everyone. They just are.

Mountain rivers are liquid expressways seeking the sea. Cold, and in most places fast, their churning passage over rock bottoms is a dare. An inflatable flotilla on the Nantahala eases through slacker waters just before entering a gauntlet of rapids.

A backpacker is, if nothing else, a nomad in search of a better view. Few vantage points are as promising as the strategically-located firetowers of the Great Smoky Mountains National Park. Hikers have left their mobile homes out of reach of bears as they ascend to the top for a hoped-for peek through the fog.

The essence of the ever-changing mountains is captured in this stream of Stone Mountain State Park near Roaring Gap in the northern part of the state. The energy of water, the substance of rock, and the color and softness of living things are the elements which, in tangled permutations, have marked and shaped the majestic hills.

Probably no other place in the United States has witnessed the marriage of boundless vision and unlimited resources to the extent manifest in Biltmore, the summer estate of George Vanderbilt. It has been called the most beautiful house in America, and one of the most opulent in the world. It is a dream rendered in stone by the finest craftsmen of their day from a mother lode of nature's resources—the very mountains which envelop the once 400,000-acre estate. Frederick Law Olmstead, the father of landscape architecture in America, molded the grounds and immediate surroundings; Richard Morris Hunt adapted and re-created a French chateau to be the suitable center of the newly-created "barony". Vanderbilt, of course, constructed a village to house the craftsmen and artisans who built the estate.

Asheville has drawn its life from the surrounding hills. Long a tourist destination, the city thrives on its heritage as the only metropolis-in-the-mountains. Today it is surging with the growth and energy of the rediscovery of high country living at its finest.

A Paisley hillside swallows the Blue Ridge Parkway which drives directly into the heartrock of this lumbering slope. In its western passage to the Great Smoky Mountains National Park, the roadway slices through ridges like an asphalt thread suturing view after splendid view.

Such simple displays, or similar ones, slow thousands of travelers at countless locations along mountain roads, as here along Highway 221 near Linville. These Mason Jars-O'Gold are the advertising and the merchandise together—packaging is everything. If the sun doesn't illuminate the languid richness, then no sales pitch at all will persuade a reluctant customer.

Skilled players tame fire and fear when they recreate the saga of the American Revolution as it was fought and won in the southern Appalachians. "Horn in the West" has provided outstanding summer entertainment since 1952.

The western mountains were a seat of the Cherokee culture and nation, and outdoor dramas such as "Horn in the West" tell a part of the story of these native Americans.

The zero visibility of fog creates a paradoxical condition for fire tower personnel near Sapphire—it's too wet to burn, which is good; but it is also too dense to maintain vigil, or work, which means a rainy day off. Just swell.

Grandfather Mountain for all seasons! In winter, rime ice is that ghostly covering which whitens and greys a world of greens, browns and blacks. Chillingly beautiful, the shimmering coating radiates a prismatic beauty with any touch of sunlight—a lasting shimmer before the shower of melting.

This languid pool is the cutting edge of one of the most impressive notches carved into the land of North Carolina. The Linville River has backed its way through solid rock leaving behind Linville Gorge. Millions of years of ceaseless plummetting into a pool have made this spectacular canyon.

Little light at all penetrates the parasols of rosebay rhododendron at Stone Mountain State Park. The large wide leaves of the plant, variously called laurel or bay, can layer so effectively that gentle rainfalls can't penetrate, and the shade they make is both patterned and deep.

At Whitewater Falls a small fern tenaciously clings to the side of a stream-honed boulder, a beautiful toehold against destruction. The plant's roots will expand, water will freeze in the hairline cracks, the rock will flake and chip, and the gristmill of the stream will carry the particles seaward.

A harvest barn inside and out. Sheltered by the rattling sere leaves of an adjacent tree, a barnful of burley tobacco cures in the cool autumn air. While the colors of the two leaves match, the riches they represent are worlds of value apart.

Dried and awaiting their turn in the mill, these oversize ears of corn were grown for either silage or meal. Winebarger's Mill, on Meat Camp Creek near Boone, will turn them into the powdery staple for future meals.

Mountain church congregations, in Valle Crucis and elsewhere, form communities: their buildings are the present; their neatened graveyards are the past; and the hills around are both the future and the forever.

Overleaf: Grandfather Mountain and the surrounding graceful elevations are more Canadian than Carolinian in winter's climates. The weather here will test the senses and soul. On this day the cold is deafening and eye-watering for an instant—before the tears freeze.

The utilitarian life history of the black locust tree is captured in this photograph—from tree to fence and back to tree. This durable wood grows where split rails, shakes and posts are needed. Untamed in shape of limb and vigor of growth, black locust has a practical purpose: to tame the mountains with the traditional products rendered from its splitting.

An early canoeist leaves a silver wake in his dawn glide across Lake Toxaway, a beautifully developed, privately-owned irregular pool of more than 640 acres of clear, drinkable mountain water.

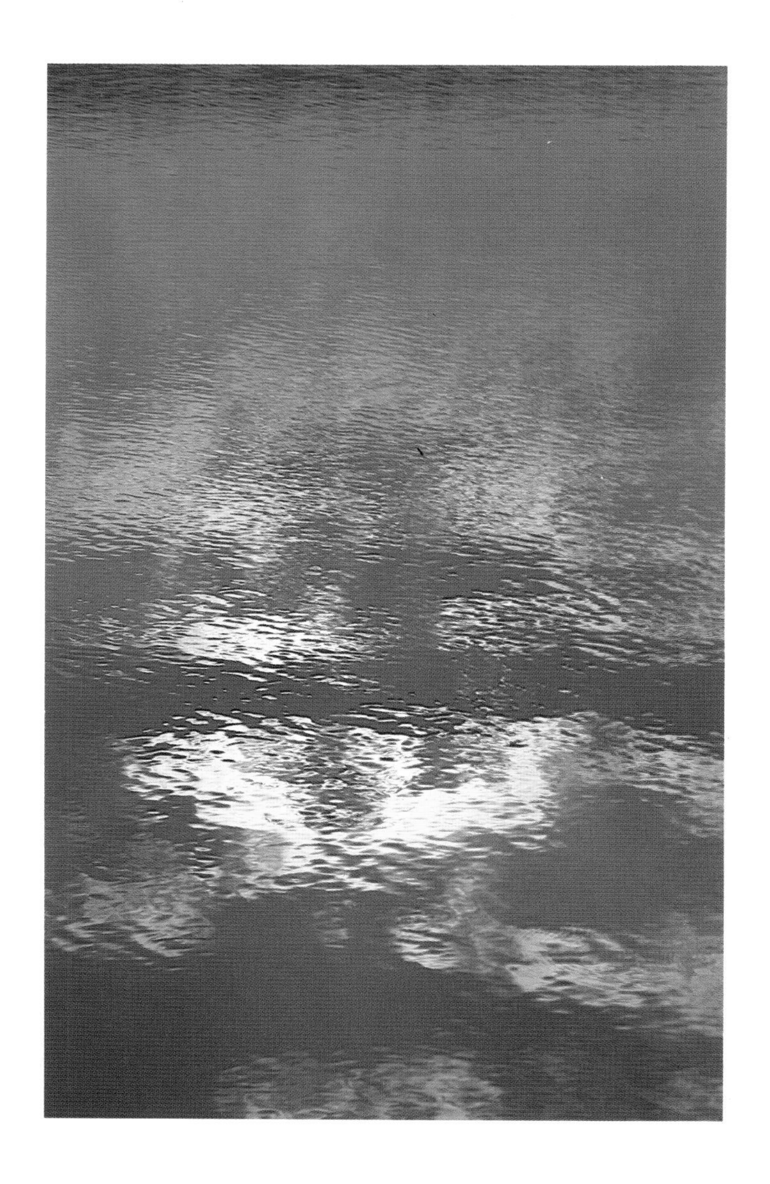

Light dances on the breeze-born facets of a pond like the reflection from a beaten copper pot. The effect is a sparkling mood of water and light, lively, airy, and a marked contrast to the mirror-like stillness of earlier hours.

Brinegar Cabin, off the Blue Ridge Parkway at Doughton Park, is a testimony to mountain practicality in building; milled lumber was expensive and used only for siding and smooth walls. Posts, however, merely needed to hold aloft a roof—neither bark nor bend mattered.

Hendersonville's climate is peculiarly predictable and perfect for apple growing. The annual Apple Festival has flourished alongside the area's emergence as a major apple-growing region. The fruit is as crisp as the air which bids buyers "welcome".

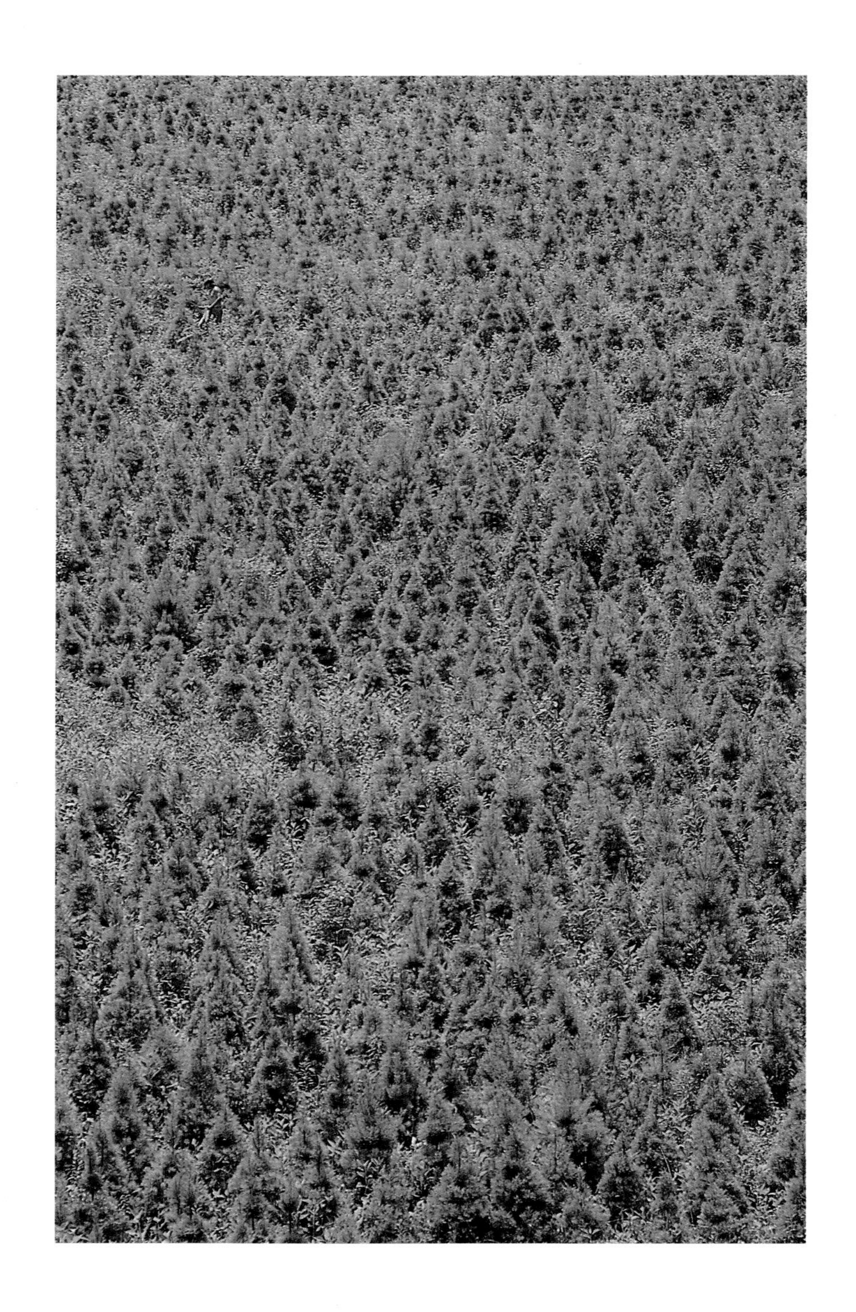

Lost in a seasonal sea of green, a grower inspects his crop of trees. Christmas tree farming has become a major mountain crop for two simple reasons: the climate is right, and the land doesn't have to be flat. It is estimated that each tree is visited one hundred fifty times by the grower before ever heading towards a treelot.

Joyce Kilmer never saw the forest his poem "Trees" inspired. The pristine, never-timbered watershed west of Robbinsville has a leafy roof that is unmatched in the state for its specimen trees, such as these twin tulip poplars.

The whitetail deer is a winsome creature, at home in his protected area at Grandfather Mountain. The deer's tawny coats allow them to merge into the woods like fog, or, because visitors only watch, they may stand boldly on display, their antlers still cloaked with a velvet covering.

The cables and catwalk of the swinging bridge at Grandfather Mountain converge in clouds, swallowing a visitor in its midst. Each year thousands cross the narrow bridge, marvel at the skill and daring of hang gliders, and photograph year-round inhabitants like Mildred the Bear.

This child is heir to a story both noble and tragic. The Qualla Boundary or Reservation is the home of the Eastern Cherokee nation, descendants of those so brave that they fled to the mountains and refused the humiliation of the Trail of Tears.

Cherokee artifacts and replicas of their daily life may be seen at the town bearing their name in the Great Smoky Mountain National Park, in the heart of the Qualla Boundary. A peaceful people, the Cherokee worked and lived with the land fashioning useful everyday products from nature, such as this gourd dipper.

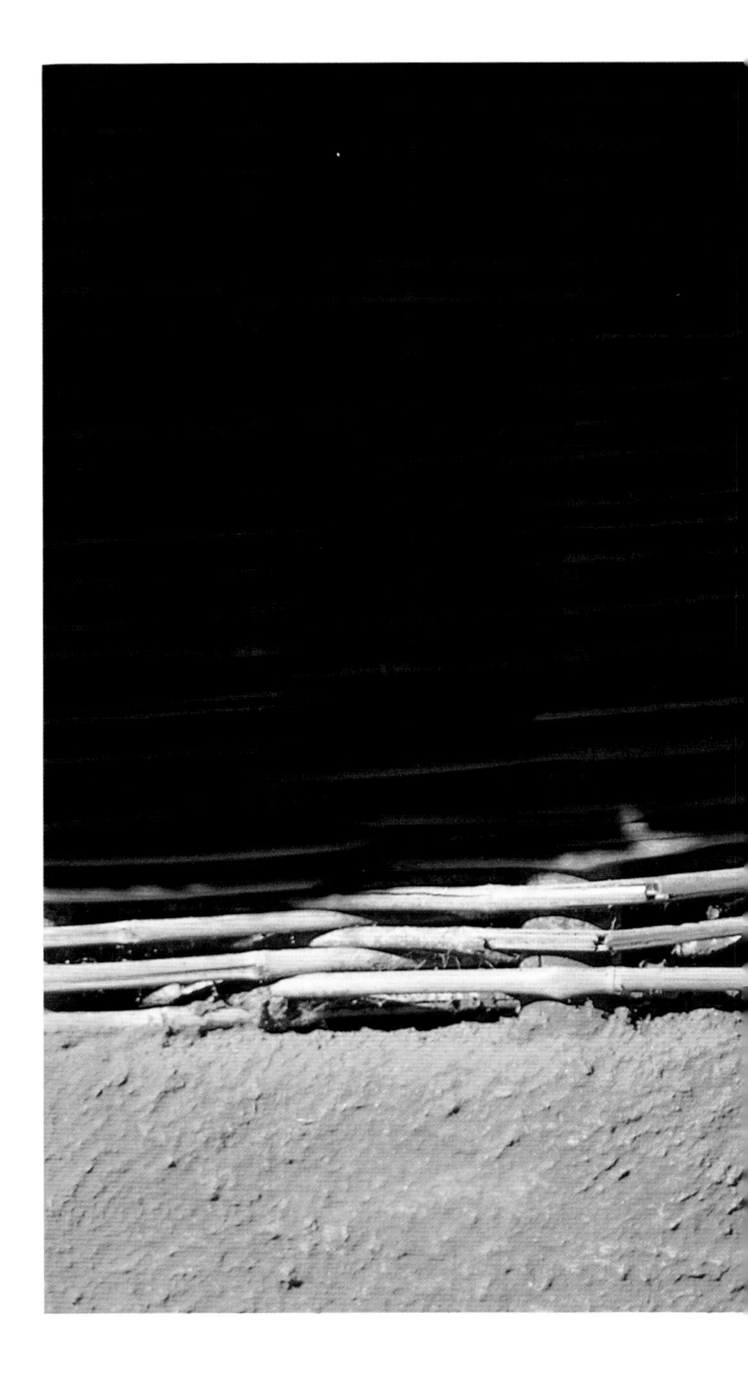

A single lacquered red tree stands sentinel in a wash of golds and yellows. In time each tree sheds its productive growing-season green to reveal fall's color.

Flour dust settles over the machinery of Winebarger's Mill, an airborne by-product of the mechanically-rendered product. The purity of the coatings' color tells an important story about the mill—business is sound, harvest has been steady. This is not the covering of abandonment, but the dust of success.

The simple lines of this horse-drawn hay rake seem etched into the pure clean backdrop of a snowfall. The shape of this implement befits its function, raking hay—this is all the tool that was needed in simpler times. One is left wondering if the same snowfall on a modern-day machine would boast as much elegance.

In contrast to the discordant rage of most waterfalls, this one showers in a diaphanous curtain. It streams free from the overhanging lip of hard cap rock, surrendering to gravity's capture in a stair-step descent.

Nature can create a complex scene and expressive mood with so few colors: a bristle of charcoal gray; a wash of blues, each tinted with subtle colors. Rich enough? Not so, for the Master is not finished until the lake waters are blushed with rose and magenta—just enough to separate the real from the mirror image.

Roan Mountain's rhododendrons beckon as no other natural botanical curiosity in the state. The cloud-swept slopes straddling the border with Tennessee are awash in pink when the bountiful trusses of the rhododendron bloom. This solitary plant fairly glows in the misty sunlight of early morning.

The complex shapes and textures of the high elevation plants of Grandfather Mountain create a landscape of stunning appearance. There is extravagant variety in leaf width and color, enough so as to impart the feeling of carefully cultivated gardens. Nature throws in the favorite color of the upland slopes, soft pinks, with flower shapes all their own.

It is nearly ritualistic to bid a mountain day farewell and it seems that no sunsets are as enticing as those which settle gently behind a distant ridgeline such as the horizon near

The French Broad River makes a languid long passage through the mountains. Its route

At Winebarger's Mill, the milling and weighing are not an act of commerce but an act of trade and trust. There is a fine line difference here—when a man's table and clothes are powdered with his product he trades not only with his wares but with his name.

The mill represents a water cycle. The miller borrows water to grind that which was grown by rainfall. Traditionally, pay was a portion of the product to which he must add water to provide sustenance for himself and his family. The mill building kept water off the entire process so it could go around again. Still does.

It is the seemingly innocent streamlets which do the yeoman's work of weathering the mountains. Particle by particle these small flows carry the mountains away. The task seems impossible, but the time needed—forever—is there and guarantees success.

Is it not fall everywhere at once? Yet individual mountain slopes are as uniformly colorful as a patchwork quilt—some subdued, some fairly howling with color. Water, temperature, sunlight and tree type determine the ultimate color of an autumn hillside, and provide the variation noticeable around any bend. Pity the hillside covered in pines; it is forever slated to be the backdrop for the shows of other tree types.

The narrow-gauge steam trains which hauled lumber and passengers through the mountains are all gone, save one. At Tweetsie, you can ride on the past through a bygone era. This railway chugs around one of the more popular visitor attractions of the Blue Ridge between Boone and Blowing Rock.

Adorned in full Highland regalia, this participant at Grandfather Mountain's Highland Games salutes the country of his forbearers. The kinship exhibited by those attending is as tightly knit as their tartans.

Pontoon boats await crews at Fontana Lake. The lake is created by the highest dam in the eastern United States and forms the southern boundary of the Great Smokies. The broad, glassy lake is used as a source of water for hydroelectric power, while the surface waters are used for recreation by guests at nearby Fontana Village.

A sliver of smooth in a washboard of ridgelines, Lake Toxaway lies like a mirror beneath the summits of containing landforms, reflecting their pictures to distant eyes.

A glistening curl of wave emerges from the deep shadows of Great Smoky Mountains

The winding two-lane roads of the mountains are part of the countryside's charm: you can only expect the unexpected around the next bend. An exception to this rule is the open air market of a local farmer. Crates of apples, cabbages, rows of honey and homemade preserves punctuate widened spots along the road curves.

Brinegar's Cabin, bounded by fence and framed by color, nestles in one of the "gaps" between ridgelines. The gaps were naturally occurring low areas along the manifold ridges and, as in this case, offered water and shelter from more severe weather. Today, the cabin is the site of a living history exhibit of traditional life in the mountains.

What is merely bold in warm weather becomes fiercely intimidating when draped by a snowcap. No longer are there the muting hues of sky and vegetation to soften such forms; everything is cast in iron-hard darks. To see rocks and trees in black and white shakes your sensibilities because it brings you face to face with the mountain-shaping forces of nature.

Snow makes geometry and calligraphy from fencelines and treelines. This is not a stack rail fence but a zigging, zagging scar across a cloth of white. Beyond, the filigree of tree limbs dusted with snow spreads like a sea-fan rendering against a grey canvas sky.

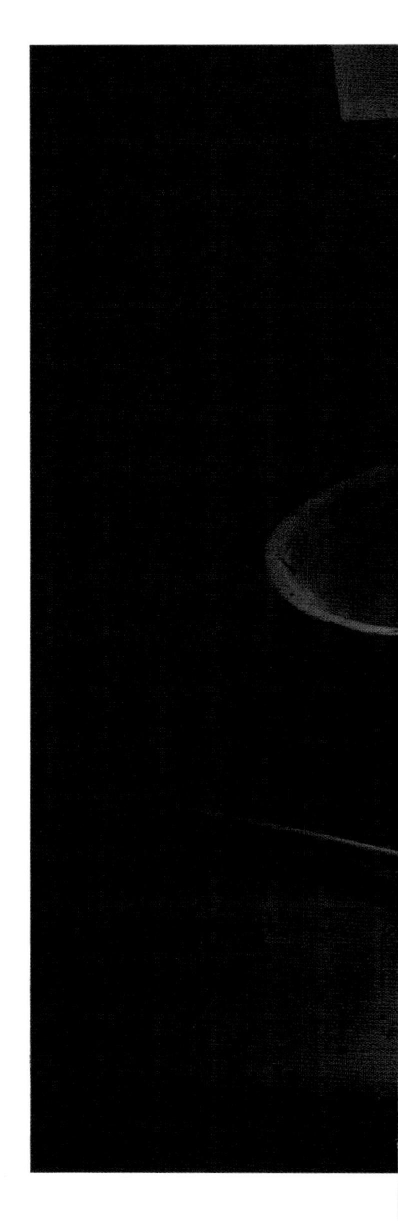

From this humble home near Asheville emerged a giant of a Governor and one of North Carolina's most renowned native sons, Zebulon Baird Vance. Leader of the state during the Civil War, Vance and his brothers left an indelible mark on North Carolina politics, now chronicled at this state historic site.

Self-reliance and inner resolve are characteristics of mountain natives. Vance, through diligence and perseverance spent at this desk, grew larger in legend and stood almost as tall as the mountains which surrounded his study.

In a visual paradox, the finite number of ridges seems infinite to the mind's eye. You can count them, or can you? Are you really sure of your numbers? It is this nagging indefiniteness, the ever-expanding possibilities of unknowable vales and hills that transfix the eye and send your imagination searching for an answer to "how many, how far"?

Frederick Law Olmstead, the crafter of the grounds at Biltmore, was a master at visual creation. He *imagined* this view at the turn of the century; he may have created it by planting these very trees. He certainly established its possibility by the location of the bridge. Olmstead knew the potential fecundity of native streambanks and their capacity for regenerating forest growth. How much of this is his doing and how much was his understanding of what nature would do through time? At Biltmore, a man directed the building of this bridge and probably walked away chuckling at the mystery he created.

At Graveyard Fields in Haywood County, ragged, rugged granite has been polished and smoothed by nature's great lapidary machine—water. The hard underlying rock of the mountains and its resistance to erosion is one reason that so many of North Carolina's waterfalls tumble down the face of the rock rather than free-fall in a shimmering veil.

"Unto These Hills", an outdoor drama which runs through the summer in Cherokee, tells the tragic story of the Cherokee Indians from 1540 until their relocation into Oklahoma on the devastating "Trail of Tears." With a dramatic gesture, a summer player brings to life the pathos and plight of the once-proud protectors of the mountains of North Carolina.

A trooper stands with rifle ready, prepared to execute the mandate of the United States Government—the wholesale removal of a people and their way of life from their ancestral grounds. As vast as the relocation effort was, it was not complete. A number of Cherokees took refuge in the mountains and refused to leave. They eventually returned to form the Eastern Cherokee nation.

There are roads in the mountains and mountain roads. The former takes you to a destination; the latter takes you on a journey through a way of life, slicing past samples of society and culture. Highway 105 between Linville and Blowing Rock is a mountain road; less traveled, but infinitely enriched by scenes such as these.

Generations to be framed by generations past . . . . The pioneer spirit which constructed this farm outside Cherokee, now a living history exhibit, was imbued with the liveliness of a child and the same love of life. Behind everything on the farm was a guiding thought: provide for those who are and for those who will be.

Below these shrouded sleepy waters is sleek excitement–trout. What trout do best is to leave the water reluctantly; what trout do second best is to refuse to place themselves in a position of having to leave the water involuntarily. Anglers love trout for these two reasons; and thereby the trout leads the angler to one of the best of all possible days—solitude and quiet on glasslike waters, trout or no.

Go ahead, take the plunge. A river raft cushions you somewhere between wet and soaked, and a successful shoot of a rapid is measured by the height of your bounce above the raft and the number who remain in the raft during the passage. The icy spray only provokes more excitement and challenge.

In the last glacial epoch, the great ice masses altered the climate of the highest elevations of the state and pushed before them the vegetation of the far north. Today these remnant forests and towering trees hang on, thriving on the cold nights and moist winds.

Watauga, Ashe and Alleghany counties are the northernmost mountain counties in the state. In these hills, sculpted by the New River and its headwater streams, are countless family enclaves—small farms settled with a self-confidence imparted by generations. This is some of the most pastoral country in the nation, an exquisite blend of wild and cultivated that communicates an abiding love of the land.

Outside the town of Brevard, October unmasks the anonymous green trees, and reveals the red and flame orange of maples, the yellow poplars, and the mighty oaks as reddened coals.

Scotland forever and—at least once a year—a Scot. The full spectacle of the gathering of clans brings fall color to MacRae Meadow at Grandfather Mountain in the middle of summer.

One of the enchanting features of the mountain maples is that they refuse to yield their color even when they finally waft to the ground. The fallen leaves of this tree are waiting in the slack shallows of a stream for a surge of water to carry them to the lowlands.

Lithe as a satin ribbon, the Oconaluftee River slides around the rocks of its basin as it flows through the ancient Cherokee homeland. This peaceful waterway draws its name from two barely pronounceable Cherokee words which meant *river* and *near* or *beside*, suitable since the Cherokee's central village was adjacent to its comforting waters.

Drop by drop, winter has sculpted chilling stalactites from the elevated raceway at Winebarger's Mill.

From underneath a snow-covered bower a passive, primitive landscape glistens. Early morning snowfall brings one of the purest of nature's moods; not only is the landscape painted with a uniform white, but it is also brushed heavily with silence.

Highway 64 west of Brevard is called the "waterfall highway" for its route skirts a number of cascades. Near Gneiss, the roadway follows the path of the Cullasaja River and offers a glimpse of its sublime waterfall.

The tenacity of life in Linville Gorge is fierce. Gnarled, broken, hollowed and weathered, this tree, perhaps a home for a nocturnal creature, still clings to life: too wounded to heal; too viable to die.

As few as ten years ago, Mount Mitchell was clad in the deep black-green of spruce and fir trees—today it is covered with skeletons. A mysterious combination of lethal elements and events is killing the rare forest of the highest peak east of the Mississippi. Bare limbs and blue sky bear mute testimony to the slow death on the mountain.

The staggering lines of fence posts stretch resolutely to a corner before turning back towards the distant ridge along State Highway 421 near Boone. Man has planted his geometry on the rolling hills of the mountains, and dropped right angles on nature. Nature drops snow on all, and suddenly, the once self-assured fenceline seems like an erratic folly.

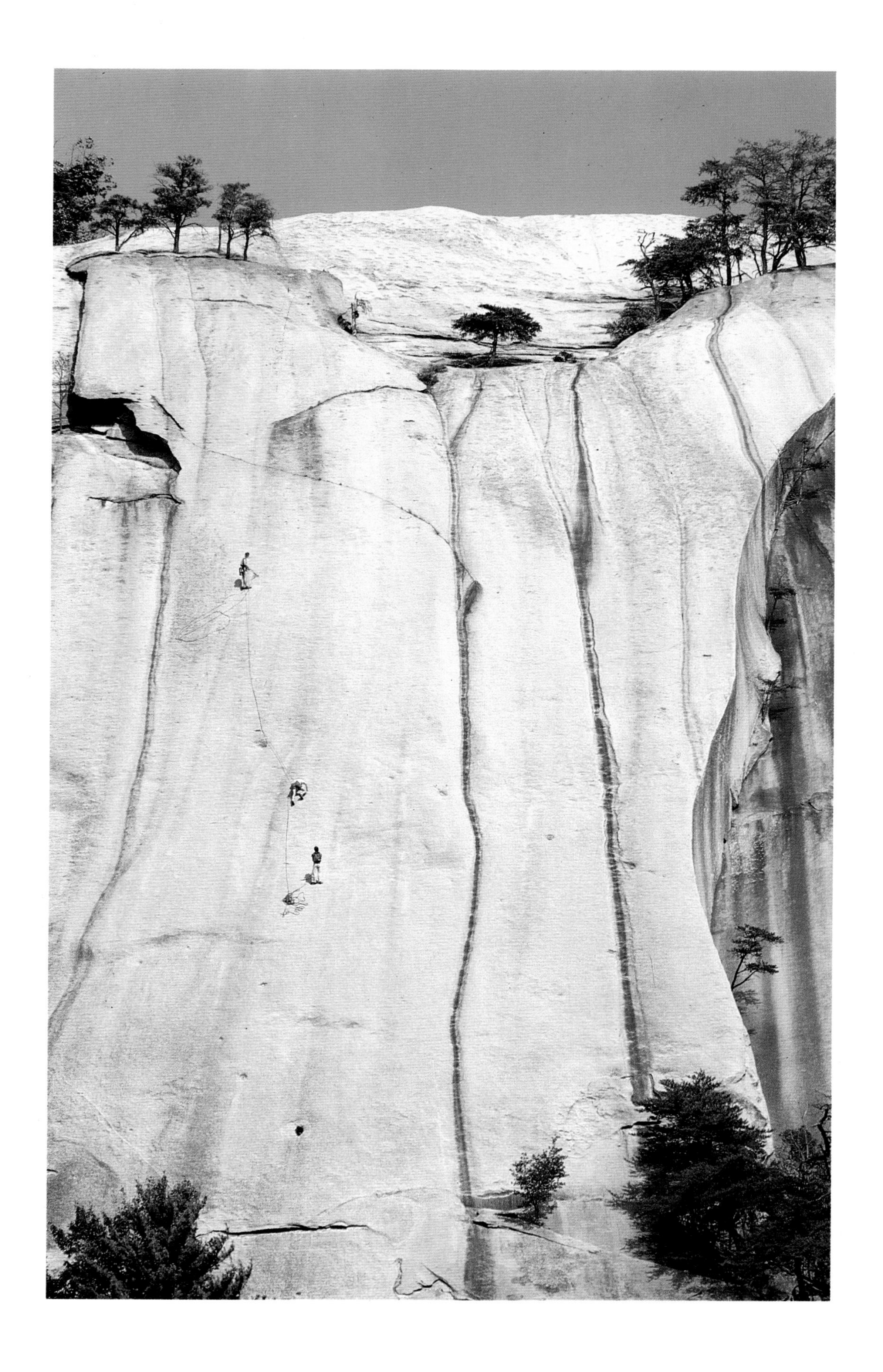

Look closely—dwarfed by the sheer immensity of its 600-foot rock face, climbers inch skyward up Stone Mountain. Enjoy this vantage point: a climber's vista on Stone Mountain is measured by the inches between minute crystalline projections laughingly called *handholds*.

To call the cows for milkin' and the hands for dinner and children to school, what better than the piercing toll of a bell. This once common communication tool has been short-circuited in its usefulness by the miles of wire-serving phones, but old lines will never age with this grace. This toller is found at Blowing Rock, at one of the many Bed-and-Breakfast inns that make the Appalachians more enjoyable.

A lush tangle of vegetation thrives just off the roadway serving the Great Smoky Mountains National Park. The most visited park in the country, this 500,000-plus acre preserve has been recommended for wilderness status.

The dark shadows of early morning frame the splendor and hope of a new day.

Swift and determined, a narrow stream is channeled, concentrated, and then released to splash on a broad fan of rock below. It has refreshed more than a few visitors traveling State Highway 221 between Linville and Blowing Rock.

A mountain water fountain . . . ahhh. There is no purer, finer drink than that which flows cleanly, clearly and coldly beside a hiking trail.

A mass of lakeside trees is mirrored in the grey waters of early morning on Lake Toxaway.

The hillsides begin to rest while a valley near Roan Mountain slowly lights to life. The twinkling below is the foxfire of civilization, soon to be matched by a starry bonanza above.